Alien Even and Alien Odd

A Math Space Adventure

Kathleen L. Stone

Enjoy these other books by Kathleen L. Stone

Penguin Place Value
A Math Adventure

Number Line Fun
Solving Number Mysteries

Riley the Robot
An Input/Output Machine

Mason the Magician
Hundreds Chart Addition

Katelyn's Fair Share Picnic
More Math Fun

Money Tree Mysteries
Adventures with Quarters

From My Quilted Heart to Yours
Heart Warming Quilts and Heart Healthy Recipes for Your Loved Ones

ISBN-13:978-1503081055
ISBN-10:1503081052

Dedication

To my wonderful husband, Gary, who has supported this crazy dream from the very beginning. I love you with all my heart!

One dark and stormy night
A light flashes across the sky,
Searching for a landing spot.
One that is safe and dry.

A tiny door opens
From the side of the pod.
Out walk two aliens
Named Even and Odd.

Alien Even decides
To head off towards the right.
Looking for things
To take back on her flight.

0, 2, 4, 6, 8
Eren

She knows if you say the number
When you count by 2's
Then it's an even number.
Let's help Alien Even choose.

1
2
3
4
5
6

First she finds *six* apples
Hanging on a tree.
Is *six* an even number?
Yes! She's as happy as can be.

1
2
3

Next she finds *three* flowers
Growing near the pod.
0, 2, 4, 6, 8
Oops! *Three* is odd!

1
2
3
4
5
6
7
8

But wait! What does she spy
Resting on those rocks?
Eight round, spotted bugs.
Quickly she puts them in her box.

Even
Only

Alien Even lugs her box
Into the pod.
Then goes back out to check on
Her brother, Alien Odd.

1
2
3
4
5
6
7

Alien Odd found *seven* spiders
During his inspection!
They'll be a great addition
To his odd collection.

1
2
3
4
5
6
7
8
9

While looking in a hollow log
He finds *nine* baby skunk.
Do you think he should add them to
The collection in his trunk?

1, 3, 5, 7, 9

Odd

1, 3, 5
7, and *9*
Nine's an odd number
So the skunk will be fine.

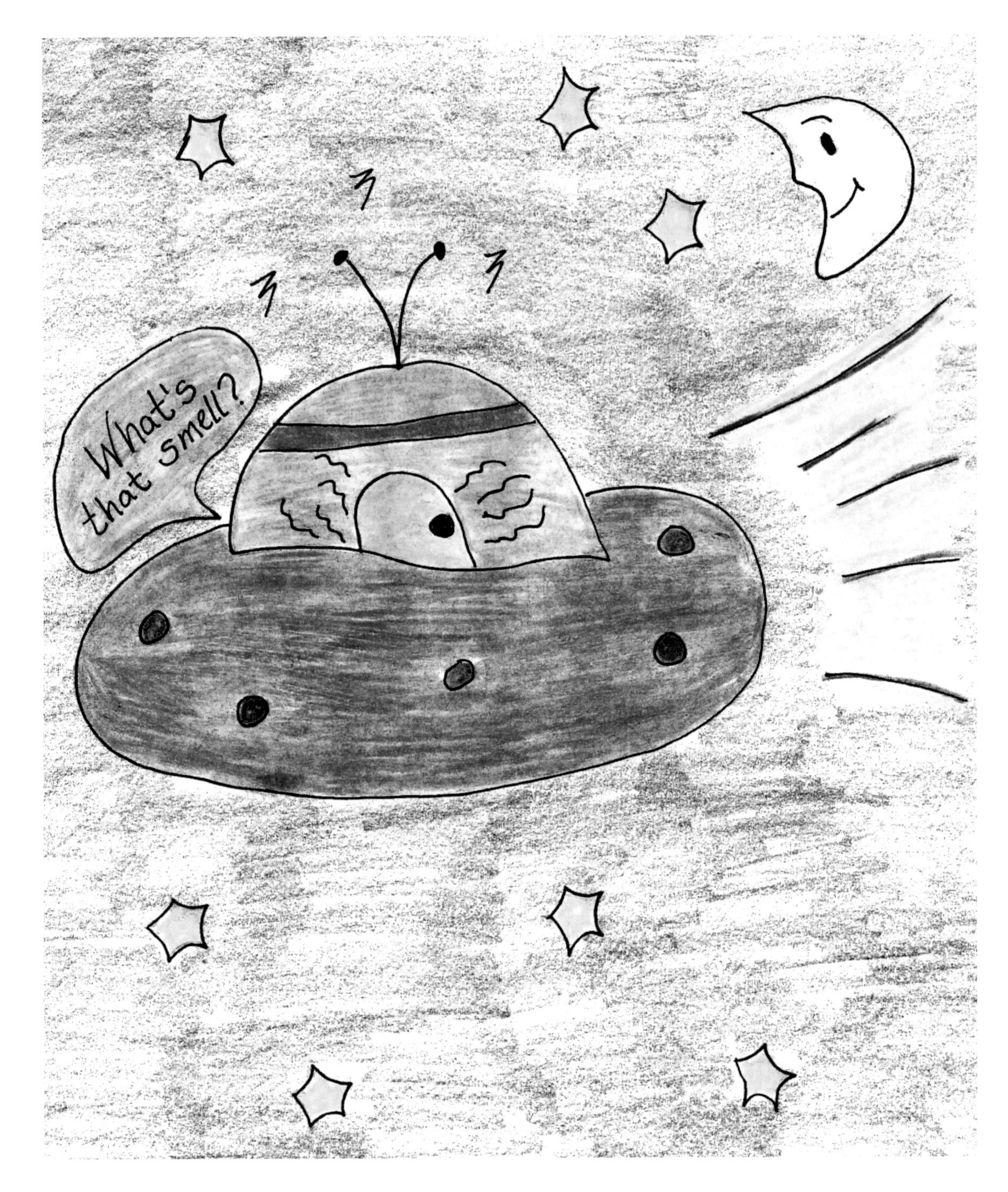
What's that smell?

Their adventure's nearly over.
Back for home they are bound.
They can't wait to show their family
All the great stuff that they found!

Even and Odd Numbers

Identifying even and odd numbers is an important number sense skill. When children work with whole numbers, it is important to know the difference between even and odd numbers. This knowledge will help them understand more complex math skills like division and prime numbers.

Even: 2, 14, 36, 128, 500

In all these numbers, the ones place has an even number. When even numbers are divided by 2, there is no remainder.

Odd: 1, 13, 65, 137, 569

In all these numbers, the ones place has an odd number. Odd numbers cannot be divided evenly.

Rules for when you add, subtract, or multiply even and odd numbers:

Adding Even and Odd Numbers	**even + even = even** 6 + 8 = 14	**even + odd = odd** 6 + 3 = 9	**odd + odd = even** 5 + 9 = 14
Subtracting Even and Odd Numbers	**even - even = even** 10 - 4 = 6	**even - odd = odd** 12 - 5 = 7	**odd - odd = even** 5 - 3 = 2
Multiplying Even and Odd Numbers	**even x even = even** 9 x 2 = 18	**even x odd = even** 4 x 5 = 20	**odd x odd = odd** 5 x 7 = 35

Enrichment Activities

Number Chants

Teach children these easy number chants …

0, 2, 4, 6, 8 … we're even numbers and that's just great!
1, 3, 5, 7, 9 … we're odd but that's just fine!

Snowball Fight

Write various numbers (both odd and even) on half sheets of paper and then crumple into balls (snowballs). Hand out to children and watch them have fun during their snowball fight. At a given signal the children stop, pick up the closest "snowball," open it and identify if it is an even or odd number!

Even or Odd Grab

Have children take turns grabbing a handful of Unifix cubes or some other items you have a lot of. Have them count their items to identify if they grabbed an even or odd amount. It's fun to see if children use the strategy of "partnering up" their items in groups of two as they are counting.

Name Game

Have children write their first name on a strip of graph paper, one letter per square. Children will place their name strip on a poster to indicate if their name is even or odd.

ABOUT THE AUTHOR

Kathleen Stone is a National Board Certified educator and is currently teaching second grade. She loves spending time with her family. She and her husband Gary live in the Olympia area. When not teaching, Kathleen can often be found quilting or sitting by the lake reading!

Math is all around us
No matter where you turn
Open your mind to the wonders of math
And all that you can learn

Made in the USA
San Bernardino, CA
10 May 2018